THE ZAMBIT RECTANGLE DREAM

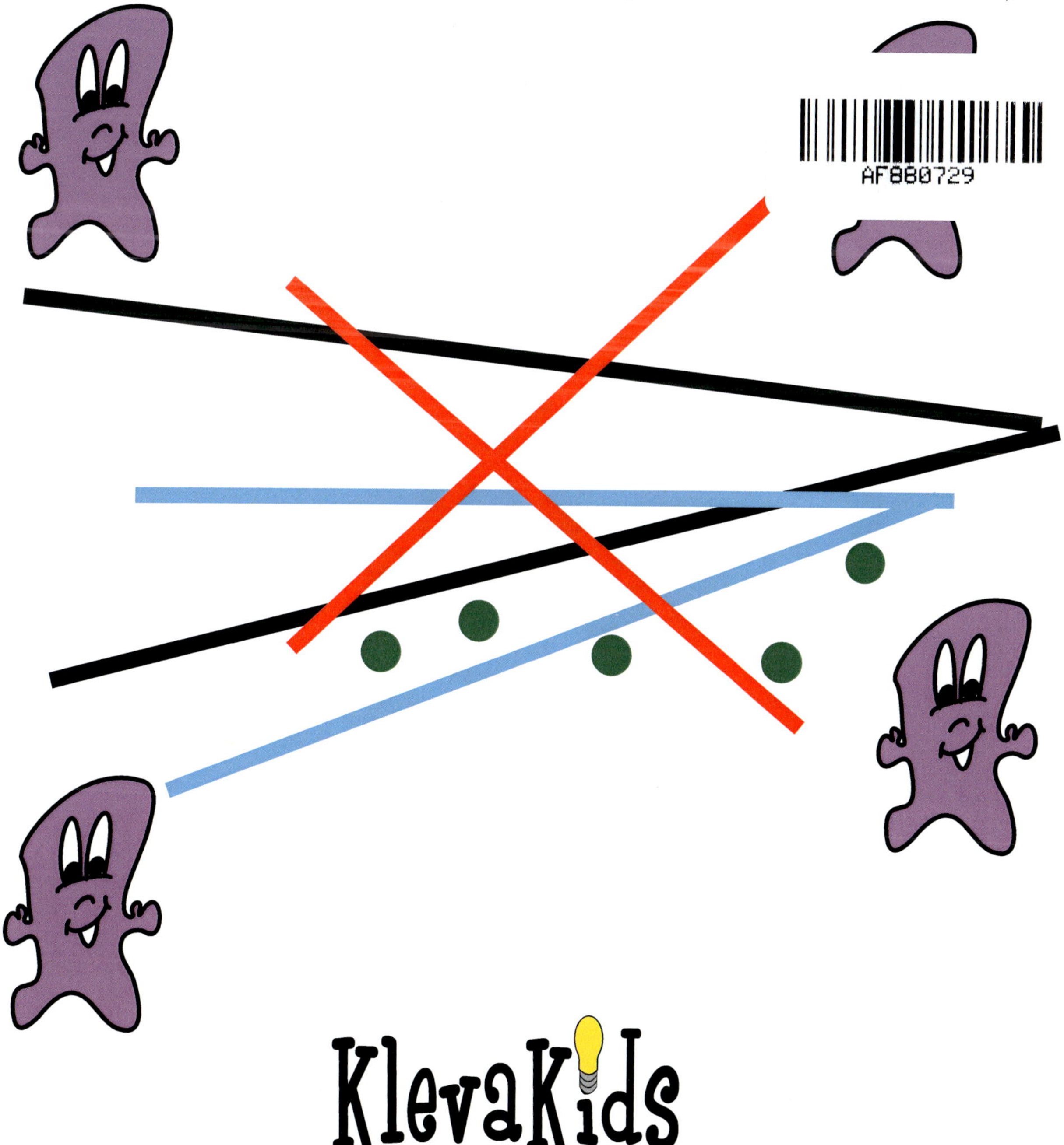

KlevaKids

Published by KlevaKids.com Inc

New York, New York

For more titles,visit our website at www.klevakids.com

KlevaKids

KlevaKids

Presents

THE ZAMBIT RECTANGLE DREAM

This story introduces the basic components of a rectangle in an enjoyable format.

The above statements are purely informational. The purpose of the book is not to teach, but to expose concepts. Read the book to a child and just allow the story to be enjoyed.

That little pile

made me so confused.

It was so messy.

I was not amused.

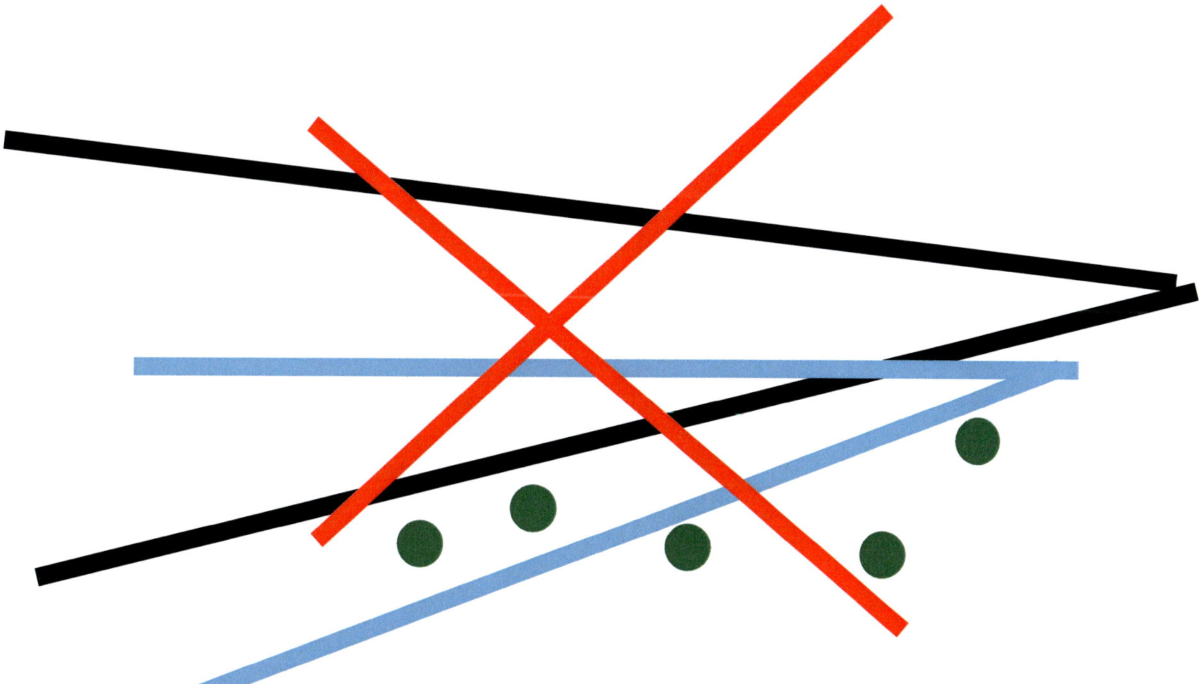

The math symbol contraption arrived

for a job it could not ignore.

Bringing order to that pile

was definitely a chore.

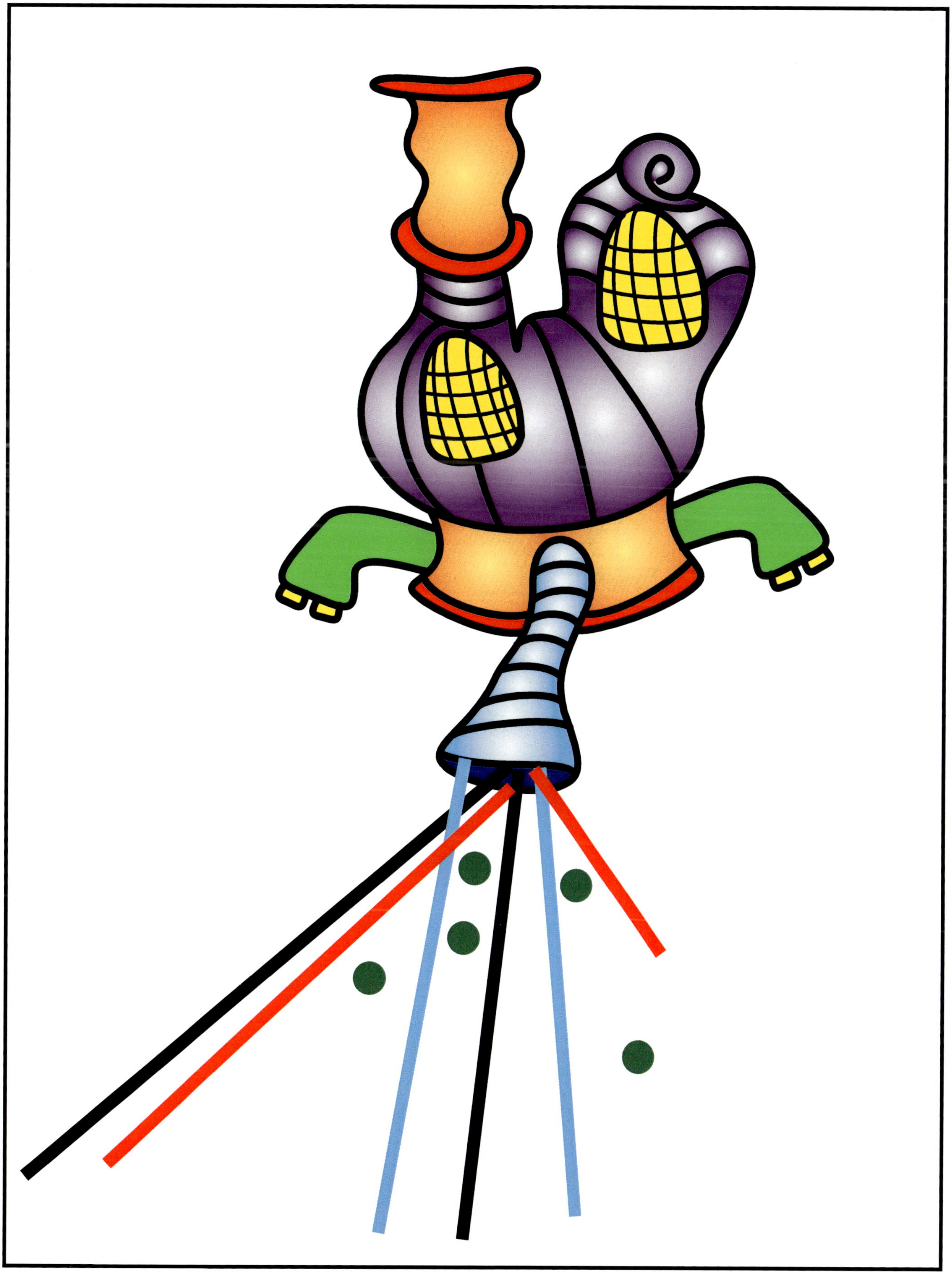

The contraption wobbled a bit.

It was strong and did not crack.

Several points then

floated from its stack.

Line segments emerged.

They were straight and true.

I had seen them before.

They were things that I knew.

The contraption then left

with everything organized.

Just line segments and points

made me quite surprised.

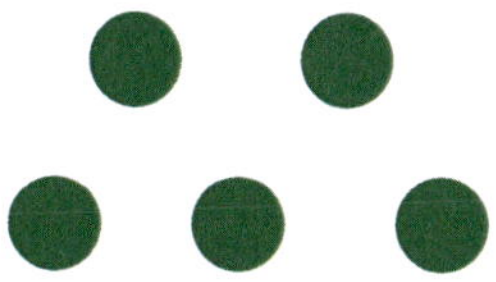

Off in the distance

something started to appear.

It was a geometry ship.

I could tell it was near.

Inside were math Zambits

and they started coming out.

Surely they could demonstrate

what this stuff was all about.

They grabbed two line segments

and made them parallel.

These were horizontal,

as far as I could tell.

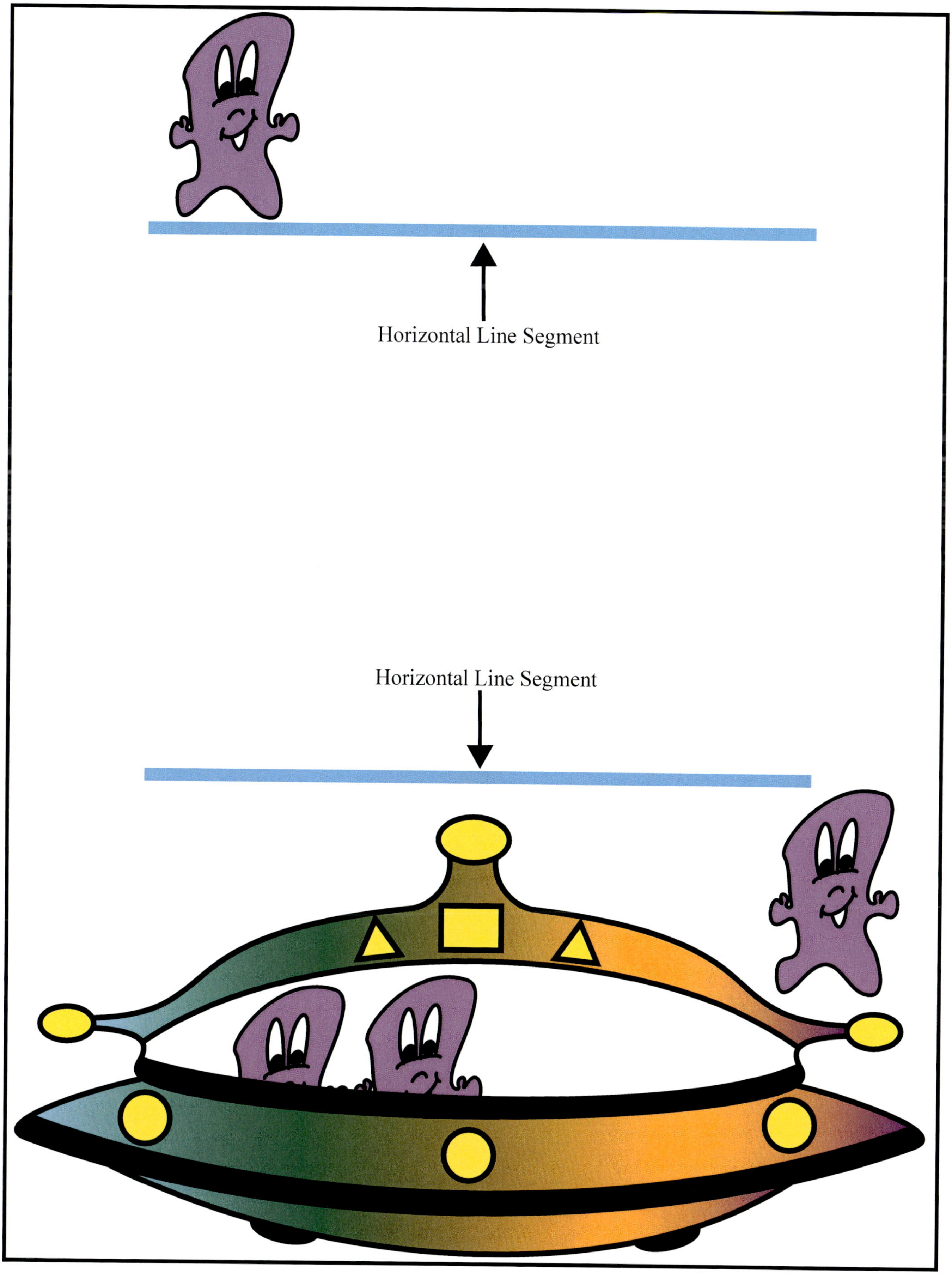
Horizontal Line Segment
Horizontal Line Segment

Two more line segments

were moved parallel with grace.

These were vertical

and going to the right place.

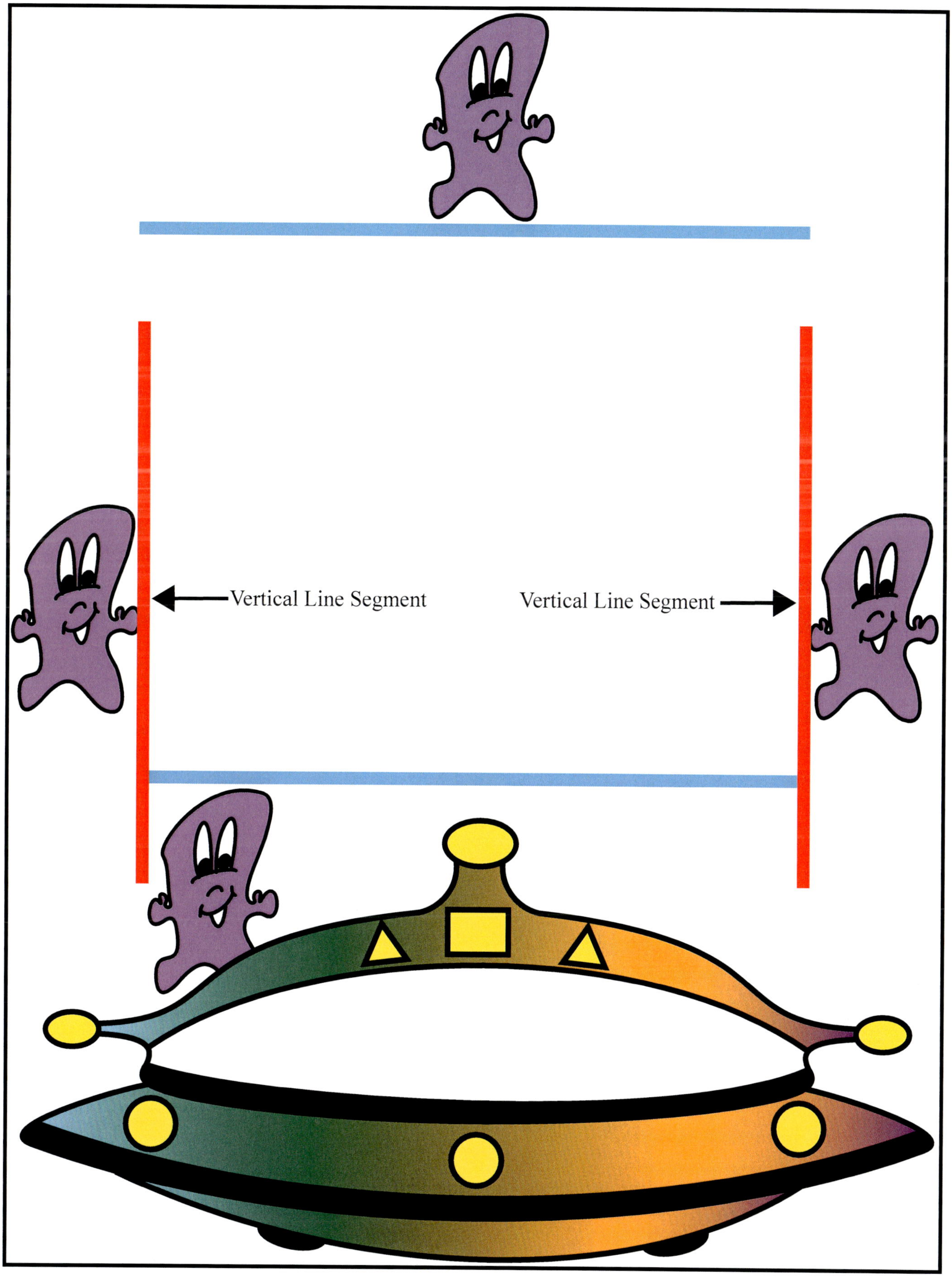
Vertical Line Segment
Vertical Line Segment

The parallel line segments

were swiftly connected.

The Zambits made a rectangle

that I clearly detected.

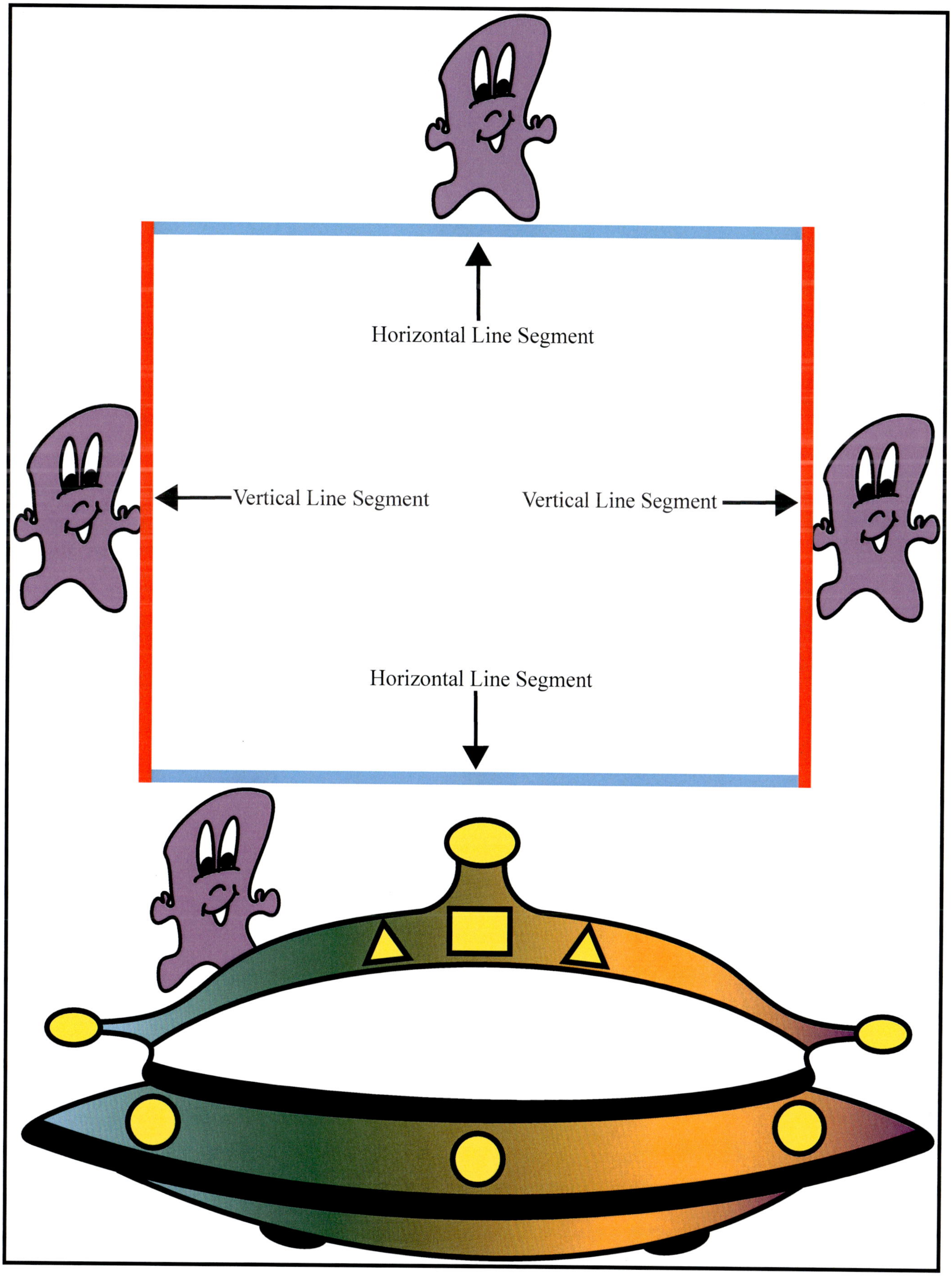

Horizontal Line Segment
Vertical Line Segment
Vertical Line Segment
Horizontal Line Segment

Each corner of the rectangle

was ninety degrees and nifty.

All of the angles added together

totaled three hundred and sixty.

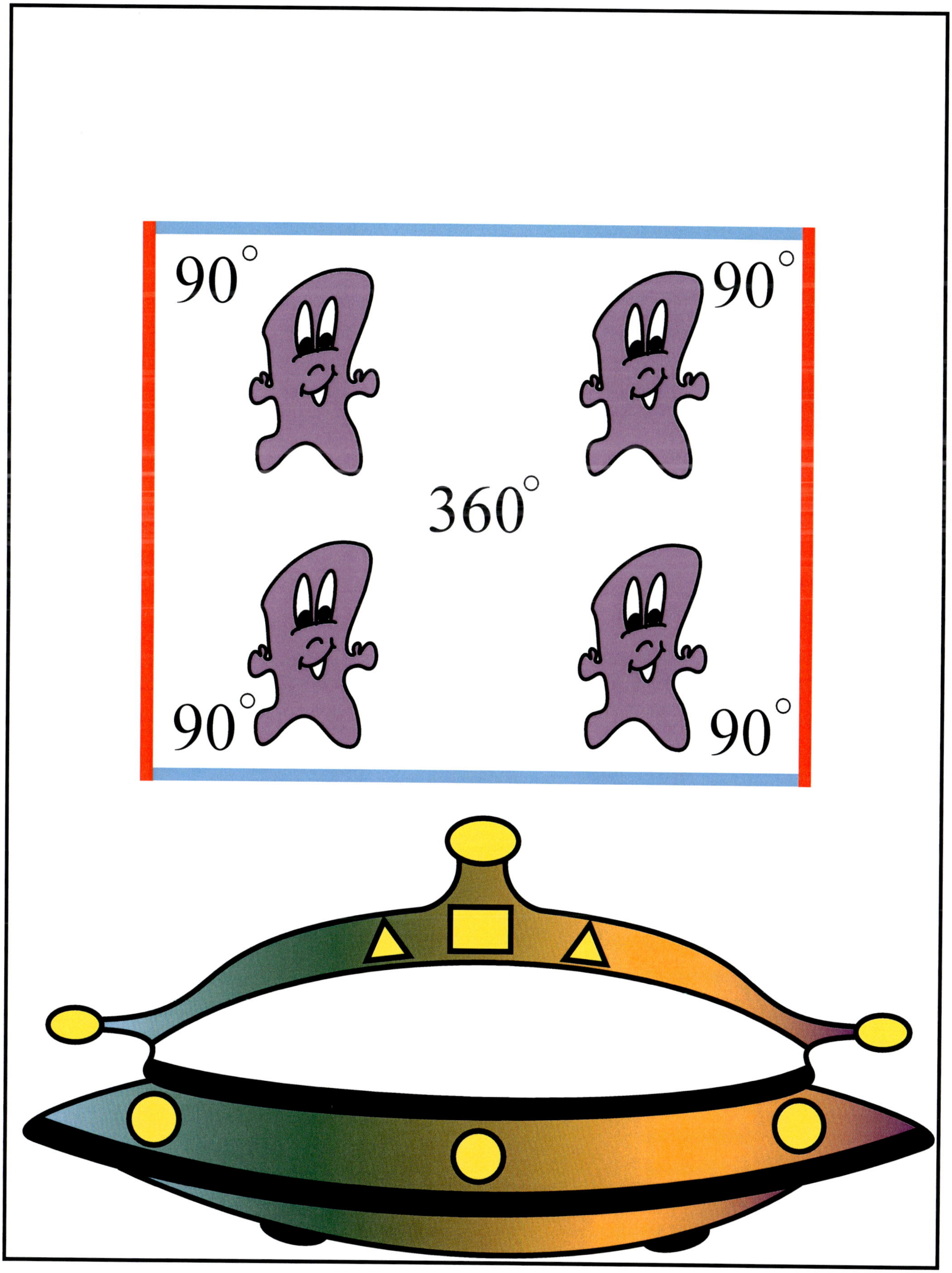
90°
90°
360°
90°
90°

Off to the corners

two Zambits went.

They created a diagonal

with a line segment.

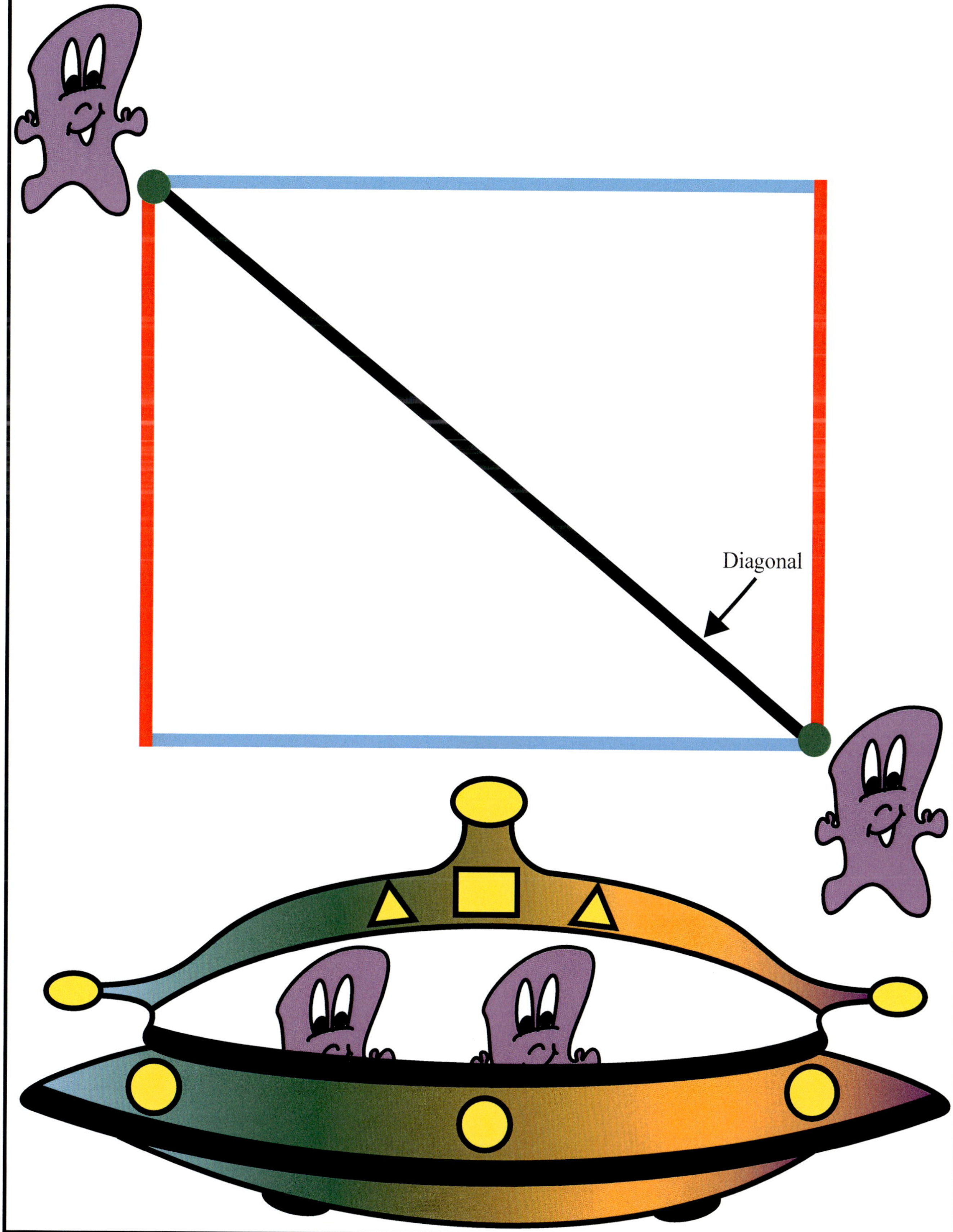
Diagonal

They created another diagonal,

quickly and very snappy.

The diagonals crossed at the center.

That made the Zambits happy.

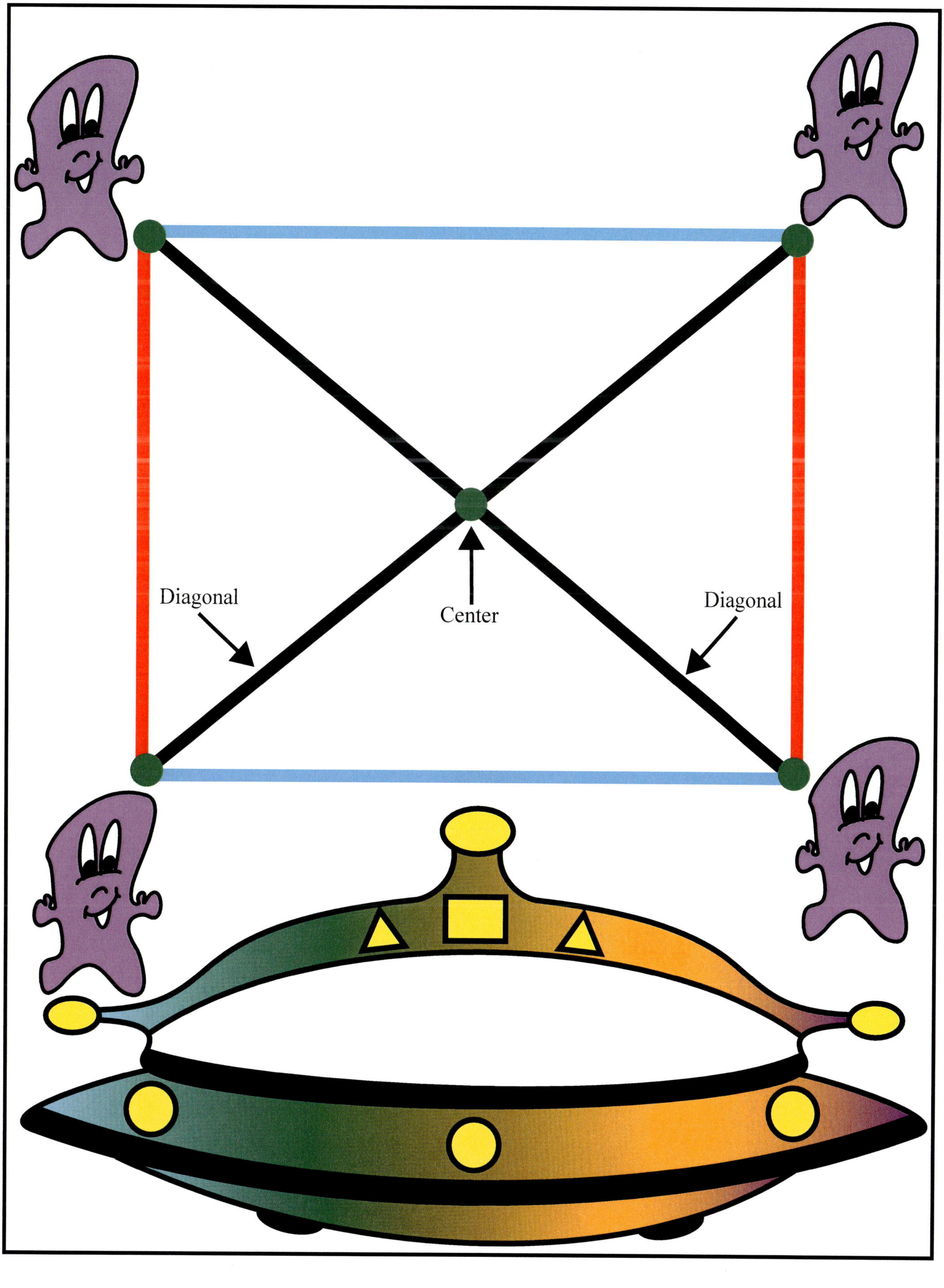
Diagonal
Center
Diagonal

The Zambits went back to their ship.

I think they had to pack.

I’ll wait until another day

when I am sure they will come back.

I love to dream of Zambits.

That's exactly what I said.

I like to keep this rectangle stuff

right inside my head.

THE END

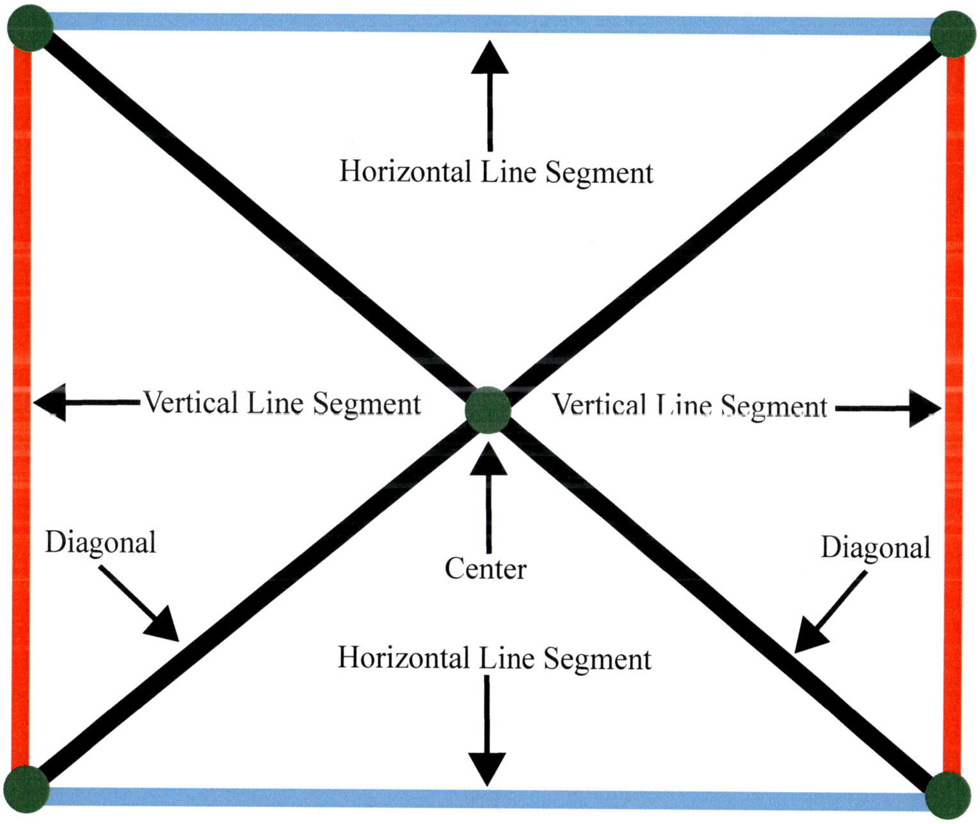
Horizontal Line Segment
Vertical Line Segment
Vertical Line Segment
Diagonal
Center
Diagonal
Horizontal Line Segment

Made in United States
Orlando, FL
04 November 2026

90203683R00021